BOOKWORMS

A Day with a
Farmer

By Katie Kawa

Cavendish
Square

New York

Published in 2021 by Cavendish Square Publishing, LLC
243 5th Avenue, Suite 136, New York, NY 10016

First Edition

Website: cavendishsq.com

This publication represents the opinions and views of the author based on his or her personal experience, knowledge, and research. The information in this book serves as a general guide only. The author and publisher have used their best efforts in preparing this book and disclaim liability rising directly or indirectly from the use and application of this book.

Cataloging-in-Publication Data

Names: Kawa, Katie.
Title: A day with a farmer / Katie Kawa.
Description: New York : Cavendish Square, 2021. | Series: Community helpers at work | Includes index.
Identifiers: ISBN 9781502658104 (pbk.) | ISBN 9781502658128 (library bound) | ISBN 9781502658111 (6 pack) | ISBN 9781502658135 (ebook)
Subjects: LCSH: Farmers–Juvenile literature. | Agriculture–Juvenile literature. | Community life–Juvenile literature.
Classification: LCC S519.K27 2021 | DDC 630.92—dc23

Editor: Katie Kawa
Copy Editor: Nathan Heidelberger
Designer: Andrea Davison-Bartolotta

The photographs in this book are used by permission and through the courtesy of: Cover Iakov Filimonov/Shutterstock.com; p. 5 Jack Frog/Shutterstock.com; p. 7 Monkey Business Images/Shutterstock.com; p. 9 Thoyod Pisanu/Shutterstock.com; p. 11 Fotokostic/ Shutterstock.com; p. 13 Praweena style/Shutterstock.com; p. 15 Syda Productions/Shutterstock.com; p. 17 bernatets photo/ Shutterstock.com; p. 19 oticki/Shutterstock.com; p. 21 Budimir Jevtic/Shutterstock.com; p. 23 Geri Lavrov/Photographer's Choice/Getty Images Plus/Getty Images.

Some of the images in this book illustrate individuals who are models. The depictions do not imply actual situations or events.

CPSIA compliance information: Batch #CS20CSQ: For further information contact Cavendish Square Publishing LLC, New York, New York, at 1-877-980-4450.

Printed in the United States of America

Find us on

CONTENTS

Caring for Crops

Farmers are very busy! They spend their days taking care of animals and plants on a farm. The plants farmers grow and care for are called crops. These crops often feed people in a farmer's **community**.

Caring for crops can be hard work. A farmer often can't do it alone. Some farms are run by families. Every person in the family helps out. Even kids can help out around the farm!

Crops come from seeds. Farmers plant seeds in the **soil**. Then, they water the soil. Seeds need water to grow into healthy plants. Farmers water their crops often. Sometimes they use machines to help them.

A farmer picks the crops when they're ready to be used. This is called harvesting. Some farmers pick their crops with their hands. Other farmers use a machine called a combine to help them harvest their crops.

Farmers grow many different kinds of plants. Some farmers grow plants for food. Other farmers grow pretty flowers. Farmers can also grow plants to make clothes. Cotton is a plant used to make clothing.

Raising Animals

Some farmers raise animals. They get up early to take care of the animals. On a dairy farm, a farmer cares for cows. The cows make milk. The farmer can get milk from the cows using a machine.

Sometimes farmers raise chickens. They get eggs from the chickens. Farmers raise other animals too. Pigs, goats, sheep, and horses often live on farms. It's a farmer's job to make sure these animals get food and water.

Working on the Farm

A farmer uses many tools and machines to get their work done. The machines can be very big! A tractor is used to pull these big machines. A farmer drives their tractor around the farm.

19

Farmers can also use computers to help them take care of the farm. Computers help them know when to plant and harvest crops. Computers can also **control** the machines that water plants and milk cows.

Farmers often sell things from their farm at a farmers' market. This is a fun place to talk to farmers! People can visit farms too. They can learn more about what farmers do.

WORDS TO KNOW

community: An area where people live; a neighborhood.

control: To direct how something works.

soil: The dirt in which plants grow.

INDEX